Musings

of a

Neurologist

A collection of papers on theoretical physics and mathematical neurology

M U S T A F A   A .   K H A N ,   M . D .

authorHOUSE°

AuthorHouse™
1663 Liberty Drive
Bloomington, IN 47403
www.authorhouse.com
Phone: 1 (800) 839-8640

Published by AuthorHouse   09/15/2015

ISBN: 978-1-5049-2530-3 (sc)
ISBN: 978-1-5049-2529-7 (e)

Library of Congress Control Number: 2015915120

<u>Dedication</u>: I would like to dedicate this book to my parents, my wife Nadia and my two sons, Sohraab and Hyder.

Acknowledgment: I would like to acknowledge my brother Mujahid Rasheed Mohammad for all his help in preparing the manuscript for this book and for all the beautiful illustrations in my previous book, "New Ideas for a New Era".

Table of Contents

The effect of Einstein's Special theory of relativity on the Universal Constants.

I) <u>Introduction</u>: Einstein's Special theory of relativity (STR) has shown us that many entities we had thought to be absolute are actually not and are relative. Entities such as time, length, mass and with this showed us the relativistic nature of our Universe. Here, we like to find out, what effect, if any, does the STR has on the Universal Constants. For this we will consider a few of the well-known and common Universal Constants. This can only be for illustrative purposes as there are just too many Universal Constants for us to consider in a single paper.

II) Here we will consider three well-known Universal Constants, (1) The speed of light in vacuum, C, (2) The Universal Gravitational Constant, G, and (3) The Planck's constant, h.

Before we delve into the details, let us first set up the stage. Following in the tradition laid by Einstein himself, let us take two inertial reference frames S and S'. Let us have S' move along the +x-axis of S at a uniform speed v, relative to S, with the +x'-axis of S' being parallel to the +x-axis of S. Let us put two physicists P and P' at the origins O and O' of S and S', respectively. These physicists will be the observers relative to whom we will be discussing the effects of STR on our Universal Constants.

(a) <u>The speed of light in vacuum, C</u>: It is a postulate of the Special theory of relativity that the speed of light in vacuum is the same relative to any inertial reference frame. This means the speed of light in vacuum, C', as measured by P' will be the same as the speed of light, C, as measured by P. In other words, we have C'=C. Since the aim of this paper is to find any effect of STR on the Universal Constants, we can say that the STR does not have any effect on the Universal Constant that is the speed of light in vacuum.

(b) <u>The Universal Gravitational Constant, G</u>: To find the effect of STR on G, let us put a spherical object of mass M' at O' that is at rest, relative to S'. This also means, $M' = M_0$, relative to S'. Now, this object has a spherical gravitational field, GF', relative to S', around it that, theoretically, extends to infinity. Since the GF' of M' extends to infinity, relative to P', it also must be spherical and extend to infinity relative to P. The gravitational field of M', relative to P, we will designate as GF. Let us put an object, μ', of unit mass, i.e. $\mu' = 1$, relative to S', at rest, relative to S', at a distance x' from O'. Now, P' will measure the Newtonian gravitational force,

$F'(M', \mu', x')$, on μ' by M' as, $F'(M', \mu', x') = -G' \frac{M'\mu'}{x'^2}$, G' is the Universal

gravitational constant relative to S'. Given the principle of relativity, our physicist P

in S will write his Newtonian gravitational force equation as $F(M, \mu, x) = -G \frac{M\mu}{x^2}$,

where G is the Universal gravitational constant relative to S and the other quantities are the equivalent, relative to S, of the corresponding quantities in $F'(M', \mu', x')$.

According to STR, we have $M = \frac{M'}{\sqrt[2]{1 - \frac{v^2}{c^2}}}$, $\mu = \frac{\mu'}{\sqrt[2]{1 - \frac{v^2}{c^2}}}$. P will notice length contraction, as

per STR, and he will measure the distance between M and ì, i.e. x, as given by the

STR, namely, $x = \left(\sqrt[2]{1 - \frac{v^2}{c^2}} \right) x'$. Substituting the expressions for M, μ and x into the

equation for $F(M, \mu, x)$, we get, after simple manipulation, $F(M, \mu, x) =$

$-\frac{G}{\left(1 - \frac{v^2}{c^2}\right)^2} \frac{M'\mu'}{x'^2}$, (equation #1). The equation #1 is the transformation of the equation

for the Newtonian gravitational force from S to S'. Again, as per the principle of relativity of STR, the form of the equations representing physical laws must be

similar in all inertial reference frames. This means, $-G' \frac{M'\mu'}{x'^2} = -\frac{G}{\left(1 - \frac{v^2}{c^2}\right)^2} \frac{M'\mu'}{x'^2}$. From

this we can conclude that, $G' = \frac{G}{\left(1 - \frac{v^2}{c^2}\right)^2}$. This is not what we had expected for a

supposedly Universal Constant. One can easily see the tremendous and profound consequences of the above relationship between G and G'. However, for now we will defer further discussion for later and look at our last Universal Constant, the Planck's constant.

(c) <u>The Planck's constant, h</u>: Let us have an object of mass m', relative to S', traveling at a uniform speed u', relative to S', along the +x'-axis. Our physicist P' will write the De Broglie equation for this object as $\lambda' = \frac{h'}{m'u'}$, where $\ddot{e}'$ is the wavelength of the object and h' is the Planck's constant relative to S'. For this same object, our physicist P will write his De Broglie equation as $\lambda = \frac{h}{mu}$. From STR we know the relation between λ and λ'. It is nothing else than the length contraction equation, because λ and λ' are lengths or distances between two consecutive peaks or troughs of the De Broglie wave of the object we are considering. Using the same argument we used in the previous section on the gravitational constant to obtain the relation between x and x', we get, $\lambda = \left(\sqrt[2]{1 - \frac{v^2}{c^2}} \right) \lambda'$. Substituting the expressions for λ and λ', respectively, we get, $\frac{h}{mu} = \left(\sqrt[2]{1 - \frac{v^2}{c^2}} \right) \frac{h'}{m'u'}$ (equation#2). Since equation #2 has to be valid for all m, u, m', u' and again as per the principle of relativity, which is the basis of the STR, the form of the equations representing physical laws must be similar in all inertial reference frames, we must conclude that $h = \left(\sqrt[2]{1 - \frac{v^2}{c^2}} \right) h'$, or

$h' = \frac{h}{\left(\sqrt[2]{1 - \frac{v^2}{c^2}} \right)}$. Here again, we see that a supposed Universal constant is not constant universally! In the discussion section, we will look at the consequences of this result also.

III) <u>Discussion</u>: In the above section we have looked at the effect of Einstein's Special theory of relativity on some common Universal Constants. We have found that if we accept the relativistic nature of the Universe, as per the Special theory of relativity (STR), then we must also accept the conclusion that some of our well-known Universal Constants cannot be constant universally. Let us now discuss the consequences of what we have found.

(a) The speed of light in vacuum, C: For this Universal Constant we found C'=C and that the STR has no effect on C. Therefore, there is no need for any further discussion regarding C.

(b) The Universal gravitational constant, G: Regarding this Universal Constant we found that, $G' = \dfrac{G}{\left(1-\frac{v^2}{c^2}\right)^2}$. Thus, the STR has a profound effect on G and that G cannot be constant universally. We can easily see the following consequences: (1) For $v \ll c$, G'=G as we would expect. (2) A moving object has a stronger gravitational field simply due to its motion compared to its gravitational field at rest. (3) The gravitational effect of the Sun on the planet Mercury is higher than its effect on Earth not just because Mercury is closer to the Sun, but also because the relative motion of the Sun is greater relative to Mercury than its motion relative to Earth. (3) For a photon we have v=c, which makes $G' = \infty$. This leads to the conclusion that a photon cannot have a gravitational field or, in other words, pure energy cannot have a gravitational field. This is not surprising since a photon does not have an inertial mass to produce a gravitational field. (4) In the gravitational field equation of Einstein's General theory of relativity, we need to replace the G by G' for non-terrestrial objects, such as our Sun. (5) We can speculate that perhaps it is the non-constancy of G that is responsible for the phenomenon we have so far attributed to the mysterious Dark Matter and Dark Energy. (6) One can easily see that there has to be innumerable sub-atomic black holes all around us simply due to the speed with which they are traveling, just as Stephen Hawking had predicted before. (7) There are many other consequences that one can derive which I am going to have to leave for the reader.

(c) The Planck's constant, h: Here also we have found the profound effect of STR on this supposed Universal Constant. We saw that according to STR, $h' = \dfrac{h}{\left(\sqrt[2]{1-\frac{v^2}{c^2}}\right)}$. The following consequences can be derived from the above expression: (1) For $v \ll c$, we get h'=h, which is as expected. (2) We see that $h' \to \infty$ as $v \to c$. This means if one is traveling in a spaceship at a speed that is very close to C, then, according to Heisenberg's uncertainty principle for position and momentum, the uncertainty of the position and momentum of a particle should increase compared to what we determine for the same particle on Earth. (3) In the Schrodinger's wave equation and other equations describing quantum mechanical processes, we will need to replace the h by h' when using those equations for non-terrestrial objects, such as the Sun or near the Event Horizon of a black hole.

IV) <u>Testing the theory and final thoughts</u>: As we have been able to test some of the results of Einstein's Special theory of relativity, it would be very helpful if we can test these results of the theory also. One way to test the non-constancy of G is using our particle accelerators. One can take a collection of identical particles, say protons, and have them in a spherical formation. We then gradually accelerate our spherically distributed protons to near the speed of light in vacuum, i.e. C, and see if the protons come closer to each other throughout and not just in the direction of motion, which is due to the length contraction effect of the STR. We then slow them down and see if the particles move away from each other to the original spherical distribution. In this way, we can get an indirect evidence for the non-constancy of G. We can also use the General theory of relativity and substitute G' for G in the gravitational field equation and observe for cosmic phenomenon that support the non-constancy of G.

To test the non-constancy of h, we can do experiments on the International Space Station or observe cosmic phenomenon that can best be explained using a value of h the is different that what we know it to be here on Earth. One can also do experiments involving our Sun and see if using a value of h that is greater by a factor

of $\left(\sqrt[2]{1 - \frac{v^2}{c^2}} \right)^{-1}$, where v is the apparent speed of the Sun relative to Earth,

compared to the value of h on Earth, best explains our experimental results. Finally, the most important point of this paper is to show that if we accept that our Universe is relativistic as described by Einstein's Special theory of relativity, then we should also be ready to accept that not all Universal Constants can be constant universally.

On some of the defects in Einstein's Special Theory of Relativity.

For the sake of keeping this a short paper, I will forgo the usual introduction and go directly to some of the gross defects present in Einstein's Special Theory of Relativity (STR). I will assume that the reader is familiar with the English translation of Einstein's original paper that was published in Annalen der Physik in 1905.

I) <u>A problem in the derivation of the equation</u>, $t' = \left(t - \frac{xv}{c^2}\right) \Big/ {}^2\sqrt{1 - v^2/c^2}$:

We will start with the following equation from Einstein's original paper on the STR:

(a) $\frac{1}{2}\left[\tau(0,0,0,t) + \tau\left(0,0,0,t + \frac{x'}{c-v} + \frac{x'}{c+v}\right)\right] = \tau\left(x',0,0,t + \frac{x'}{c-v}\right)$. The problem that is present in this equation is the following:

Einstein says that a beam of light, starting at O' in S', reaches x' in time $\frac{x'}{c-v}$, relative to an observer in S. Similarly, he also says that the time for the same beam of light to reflect at x' and reach O', again relative to the same observer in S, is given by $\frac{x'}{c+v}$. This cannot be valid. By doing this Einstein is in effect applying the Galilean addition of velocities to the speed of light in vacuum. This clearly contradicts his postulate of the constancy of the speed of light in vacuum relative to any inertial frame of reference that may or may not be in relative motion with respect to a beam of light traveling in vacuum.

This problem in Einstein's derivation of $t' = \left(t - \frac{xv}{c^2}\right) \Big/ {}^2\sqrt{1 - v^2/c^2}$ is enough to

make the entire derivation of the equation invalid.

II) <u>Problems with the equation</u>, $t' = \left(t - \frac{xv}{c^2}\right) \Big/ {}^2\sqrt{1 - v^2/c^2}$:

The following problems can be seen with the above equation relating t' and t:

(1) Unlike for S, where we have a single time 't', that applies to the entire S, we have an infinite number of times t' for S' due to the infinite number of values for the x in the equation for t'.

(2) To derive his famous time dilation equation, $t' = t\,(^2\sqrt{1 - v^2/c^2})$, Einstein

placed a clock at O', thereby making x=vt, in $t' = \left(t - \frac{xv}{c^2}\right)\Big/ {}^2\sqrt{1 - v^2/c^2}$, and

obtained $t' = t\,(^2\sqrt{1 - v^2/c^2})$. This means, the time dilation equation, $t' =$

$t\,(^2\sqrt{1 - v^2/c^2})$, is valid only for x=vt and not generally, i.e. for $x \neq vt$.

(3) Since, both x and t are independent variables, we can see that for any given time, t, in S, there will always exist an x, in S, equal to $\frac{c^2}{v}t$, making t'=0 at that x. This means, for the observer in S, the time t' at $x = \frac{c^2}{v}t$ will be at a standstill. Not only this, we can also see that for $x > \frac{c^2}{v}t$, we get t'<0, i.e. the observer in S will notice the time t' moving backwards!

From the consideration of the above problems with the equation for t', we can conclude that using the Special Theory of Relativity one can actually determine an Absolute Inertial Frame of Reference, S, by measuring (t')'s, of any given non-absolute inertial frame of reference S', at various points x of S. This contradicts the basic assumption of the STR that there is no Absolute Inertial Frame of Reference. In other words, the STR has a built-in self-contradiction.

III) Conclusion:

The above sections detail just two of the gross defects that are present in Einstein's Special Theory of Relativity. I will not go into the many other defects that are present in the STR, for the sake of keeping this a brief paper. This year, the physics community and the world is celebrating the 110th birthday of the STR and yet all the brilliant minds throughout the world are keeping quiet concerning the most basic and blatant defects that are present within this theory. Even Einstein's General Theory of Relativity (GTR) has to be a flawed theory for the simple reason that one of it's basic assumptions is that the STR should be locally valid.

In the end, I like to say that if the human society is interested in genuine further scientific progress, then it must discard both of Einstein's relativistic theories, whose time came and went, and replace them with theories that are mathematically, logically and physically correct and self-consistent.

A relativistic theory based on the invariance of Newton's second law for motion and the constancy of the speed of light in vacuum.

I) Introduction:

Let me begin by saying that it was never my intention to develop a new relativistic theory in order to somehow supersede Einstein's Special Theory of Relativity. My only intention was to find out, if possible, a relationship between the Newtonian Time and Einstein Time. After obtaining the said relationship, I noted that I could develop a set of relativistic transformation equations between two non-Newtonian inertial reference frames, S and S' with S' moving at a constant speed, Γ, along the $+x$-$axis$ of S. These equations turned out to be quite different from the Lorentz-Einstein equations in Special Theory of Relativity and had consequences that in some cases were qualitatively similar to those of Special Theory of Relativity and others that were very different both qualitatively and quantitatively.

II) Newton's second law for motion:

I do not have to spell out what the Newton's second law of motion is, as it is quite well known. However, I will state it's mathematical formulation, which is given by, $\vec{F_0} = \frac{d\vec{P_0}}{dt_0}$. Here, (a) $\vec{F_0}$ is the Newtonian force, (b) $\vec{P_0}$ is the Newtonian momentum and (c) t_0 is the Newtonian Time. Underlying this equation are the assumptions, (1) there exists an Absolute Newtonian inertial reference frame, which we will designate by S_0 and whose coordinates are (x_0, y_0, z_0) and which is at absolute rest, and (2) there exists an Absolute Newtonian Time which is designated by t_0 associated with S_0. Now, both S_0 and t_0 are non-measurable entities and there is nothing that we can say represents them. If we take a non-Newtonian inertial reference frame S in which we use the Einstein Time, which we will represent by the letter t, and whose coordinates we represent by (x, y, z) then the Newton's second law for motion, in this frame of reference, will be given by $\vec{F} = \frac{d\vec{p}}{dt}$.

9

III) Relationship between t_0 and t:

Let us have S move along $+x$-axis of S_0 at a uniform absolute speed 'u', relative to S_0. Using our assumption of the invariance of the second law of motion we have $\frac{dP_0}{dt_0} = \frac{dP}{dt}$ (equation #1). Now, we will assume that $dt_0 = \lambda \, dt$ and $dP_0 = \lambda dP$. Integrating the time differential equation and putting the integration constant equal to zero, so that the clocks are synchronized, i.e., when $t_0 = 0$, $we\ also\ have\ t = 0$, we get, $t_0 = \lambda t$. The relationship between x_0 and x will be given by $x = x_0 - ut_0$. From this we get $\frac{dx}{dt} = \frac{dx_0}{dt} - u\frac{dt_0}{dt}$. Taking $t_0 = \lambda t$ we get $dt_0 = \lambda \, dt$ and putting it in the above equation we get $\frac{dx}{dt} = \lambda \frac{dx_0}{dt_0} - \lambda u$. Now, putting v_0 for $\frac{dx_0}{dt_0}$ and v for $\frac{dx}{dt}$, we get after simple manipulation $\lambda = \frac{v}{v_0 - u}$ (equation # 2). The equation #2 is valid for all v and v_0, including light. However, for light, we have, according to our assumption of it's constancy, $v_0 = v = c$. This leads to $\lambda = \frac{1}{(1 - u/c)}$. From this we see that the relationship between Newtonian Time and Einstein Time is given by $Newtonian\ Time = Einstein\ Time / \left(1 - \frac{u}{c}\right)$. We know that u is not measurable, which makes determination of Newtonian Time not possible, given Einstein Time.

IV) The set of relativistic transformation equations between two non-Newtonian inertial reference frames S and S':

Let us now assume that we have two non-Newtonian inertial reference frames, S and S' that are moving at uniform absolute speeds v and v' respectively relative to S_0 along the $+x_0$-$axis$. We can immediately see that, $x + vt_0 = x_0 = x' + v't_0$ and $y_0 = y' = y$, $z_0 = z' = z$. After some simple manipulation we get the following set of transformation equations from S to S': $x' = x + \left(\frac{v-v'}{c-v}\right)ct$, $y = y'$, $z = z'$ and $t' = \left(\frac{c-v'}{c-v}\right)t$.

Let us now put Γ for $\left(\frac{v'-v}{1-\frac{v}{c}}\right)$ into the above equations. This will give us the following set of Relativistic Transformation Equations from S to S': $x' = x - \Gamma t$, $y' = y$, $z' = z$ and $t' = \left(1 - \frac{\Gamma}{c}\right)t$.

From the consideration of these equations we can conclude, (1) the absolute and non-measurable quantities are eliminated, (2) the set of equations relating the coordinates (x', y', z') to (x, y, z) is exactly the same if we consider S' moving along $+x$-$axis$ of S at a uniform, relative to S, speed of Γ, (3) we can, from this point on, totally ignore the S_0 and t_0 as they have become irrelevant, (4) the speed, Γ, is a measurable speed and not something fictional, (5) for now, there is no, a priori, upper limit on the value for Γ, and (6) one can easily find out that our transformation equations give the same speed of light, in vacuum, in both S and S'.

V) Conclusions:

1) Using symmetry we get the following transformation equations from S' to S:
$x = x' + \Gamma t'$, $y = y'$, $z = z'$ and $t = \left(1 + \frac{\Gamma}{c}\right)t'$.

2) One can easily prove that our set of transformation equations are invariant between S and S'.

3) Suppose, we have a rod of length l', at rest, relative to S', and lying along the $+x'$-$axis$ of S'. Then given that the speed of light is the same in both S and S' and $\Delta t' = (1 - \frac{\Gamma}{c})\Delta t$, we get, $l' = \left(1 - \frac{\Gamma}{c}\right)l$, where, l is the length of the same rod, if it were lying, at rest, along the $+x$-axis of S. This equation is similar, qualitatively, to the length contraction concept we encounter in Einstein's Special Theory of Relativity. The length contraction equation puts an upper limit on Γ, namely, $\Gamma \leq c$, because negative l' has no physical meaning. Also, except for a photon, which has zero mass, $\Gamma < c$, as for $m \neq 0$, the l' will have an infinite density per unit length for $\Gamma = c$.

3) Taking the equation $t' = \left(1 - \frac{\Gamma}{c}\right)t$, we get $\Delta t' = \left(1 - \frac{\Gamma}{c}\right)(\Delta t)$. This is also similar to the time dilation concept we see in the Special Theory of Relativity. For light, we

have, $\Gamma = c$, which makes $\Delta t' = 0$, i.e. for light, there can only be the present moment and time "jumps" from one present moment to the next. This is not unlike the quantum jumping we encounter in many quantum phenomena. This suggests that relative to light, time is not smooth but quantized and consists of a series of "jumping" presents.

4) From our equation between the Newtonian Time and Einstein Time, we see that, at the moment of creation, i.e. Big Bang, u=0, making the Newtonian Time equal to the Einstein Time.

5) From the equation, $dP = \lambda' dP'$, where $\lambda' = \left(1 + \frac{\Gamma}{c}\right)$, and the equation, $v = \frac{v' + \Gamma}{1 + \frac{\Gamma}{c}}$, where v' is the speed of an object relative to S' and v is the speed of the same object relative to S, we obtain $m(v)(v - \frac{\Gamma}{1 + \frac{\Gamma}{c}}) = \left(1 + \frac{\Gamma}{c}\right) m'(v')v'$. We obtain the following equations from the above equations:

a) Relativistic mass, $m(v) = m_0(1 + \frac{v}{c})^2$, where m_0 is the rest mass of the object. This relativistic mass equation can also be written as, $m(v)c^2 = m_0(c + v)^2$, or $m(v)c^2 = m_0c^2\left(1 + \frac{v}{c}\right)^2$. We see that both sides of this equation have units of energy. Thus, if we represent $m(v)c^2$ by E_m, we get, $E_m = m(v)c^2$, which is the famous mass/energy equation. However, it is also clear that the relativistic mass m (v) is given by a different expression than what we are used to from the Special Theory of Relativity. We can also see that the rest energy of an object of rest mass, m_o, is given by, $E_o = m_o c^2$. Thus, we see that the famous matter/energy equation is a normal consequence of our relativistic theory.

b) Relativistic Kinetic Energy, E_K, is given by $m_0 v^2 \left(\frac{1}{2} + \frac{4v}{3c} + \frac{3v^2}{4c^2}\right)$. We see that for non-relativistic speed v, $E_K = \frac{1}{2} m_o v^2$, which is just what we would have expected.

c) The total energy, E_T, that is contained in an object with relativistic mass, $m(v)$, traveling at a relativistic speed, v, is given by: $E_T = m(v) c^2 + E_K$.

6) The muon decay paradox is easily solved with our theory as follows: using the length contraction equation we have obtained, we see that for a muon created 50km above the surface of the Earth and moving at 0.995c, the length it has to travel

reduces from 50km to just 250m, which is well within the 660m that one would expect the muon to travel without length contraction effect.

7) The "twin paradox" does not arise in our theory at all, since, in our theory, for a stationary clock in S', $\Delta t' < \Delta t$, relative to S. This is obtained from $t' = \left(1 - \frac{\Gamma}{c}\right) t$ and for a stationary clock in S, $\Delta t > \Delta t'$, relative to S'. This is obtained from $t = \left(1 + \frac{\Gamma}{c}\right) t'$.

8) The relativistic addition of speeds according to our theory is given by, $v = \frac{v' + \Gamma}{1 + \frac{\Gamma}{c}}$, from which we have:

a) For $v', \Gamma \ll c$, $v = v' + \Gamma$, which is our non-relativistic Galilean addition of speeds.

b) For just $\Gamma \ll c$, we have, $v = v' + \Gamma$. This superficially looks like the Galilean addition of speeds, but it is not, because v' can be relativistic, which is unlike the speeds in the Galilean expression. This is the physical/logical basis for our particle accelerators, where, even though $\Gamma = 0$, relative to the laboratory/Earth, v' can be relativistic, relative to the accelerator. Suppose S' is part of an object, which is being observed from S. In this case, the speed of S', relative to S, is equal to the speed of the object, relative to S and the speed of the object, relative to S', is zero. Since, v' in the above equation, can be relativistic, this means, the speed of S', under these specific conditions, can also be relativistic. These arguments can be expressed as, v'=0 and $\Gamma = v'$, which means, v=v', where v' can be relativistic and is the speed of S', relative to S. As we have been representing the speed of S', relative to S, by Γ, we get, $v = \Gamma$, where, $0 \leq \Gamma \leq c$ and $\Gamma = c$ for $m_o = 0$ only, such as a photon. This kind of inertial reference frame, i.e. one which is a part of an object, we will call S' of the first kind, because in the next section we will see that our relativistic theory shows that there exists a S' of the second kind, and though it is, generally, quite different from S' of the first kind, under certain very specific conditions, it becomes S' of the first kind.

c) We use our equation, $v = \frac{v' + \Gamma}{1 + \frac{\Gamma}{c}}$ and put v'=0, giving us $v = \frac{\Gamma}{1 + \frac{\Gamma}{c}}$. The S' here, whose speed, relative to S, is Γ, is S' of the second kind. One can easily see the difference between the two kinds of S' based upon the expression for v for each of them. We can also see that for $\Gamma \ll c$, the expression, $v = \frac{\Gamma}{1+\frac{\Gamma}{c}}$, becomes $v = \Gamma$, which is the same as for S' of the first kind with $\Gamma \ll c$. Therefore, in the non-relativistic realm, the two kinds of S' merge into one kind. This is the reason we did not noticed the existence of the second kind of S'. The relativistic mass equation, given in conclusion #5 is applicable for S' of the first kind.

d) As we have seen that S' comes in two kinds, it is very important when trying to explain any phenomenon, to be explicit regarding which kind of S' is being used, as the answer may be dependent on the kind of S'.

IV) Final thoughts and testing the theory:

One can ask as to why do we need a new relativistic theory when we already have Einstein's Special Theory of Relativity (STR), which has stood the test of time for 110 years. It is true that unless a new theory has not only something new to offer and can also explain phenomenon that are beyond the current theory, it is only a curiosity. But, I maintain that our relativistic theory goes beyond Einstein's STR, as follows:

1) From the relationship between Newtonian Time and Einstein Time, we see that at the moment of Creation/Big Bang, u was equal to zero, making the Newtonian Time equal to Einstein Time. Once, u was no longer equal to zero, then the Newtonian Time was no longer equal to the Einstein Time. This means that the Newtonian Time did not ceased to exist once u was not equal to zero, but simply became non-measurable, unlike the Einstein Time. It may be found that certain phenomenon, such as quantum entanglement, can be better or can only be explained using the Newtonian Time in the wave equation describing the quantum entanglement. I have already written a paper that shows exactly this in complete detail. Since STR has no such relation between the Newtonian Time and Einstein Time, and it says that the maximum speed with which information can travel is equal to the speed of light in vacuum, an explanation of the quantum entanglement phenomenon will be beyond STR, as it uses Einstein Time only.

2) We have already seen that the "twin paradox" in STR, which is due to,

$\Delta t' = \left(\sqrt[2]{1 - \frac{v^2}{c^2}} \right) \Delta t$ and $\Delta t = \left(\sqrt[2]{1 - \frac{v^2}{c^2}} \right) \Delta t'$, never arises in our theory.

There is no need to bring in the "ad hoc" explanation, that the spaceship carrying one of the twins, is non-inertial if we consider the entire round trip journey.

3) Let us consider the well-known muon decay paradox. As per the STR, the length contraction should be equal to 4994m from 50km, while our theory says it should be equal to 250m from 50km, for a muon traveling at 0.995c. If there is no length contraction, then the muon will only be able to travel a distance of 660m and should not be detected on the surface of the Earth, which is contrary to the observations. From above length contraction results based on STR and our theory, respectively, we see that our theory gives a distance that is well within 660m unlike the STR. In order to explain the paradox, STR uses time dilation effect on the muon decay time. This results in increasing the decay time by a factor of 10. If we use the same argument and apply to our theory, then we find that our theory gives a result of 200. Thus, our theory explains the muon paradox not just through time dilation like STR, but by both time dilation and length contraction. This means we can use, according to our theory, both time dilation effect and length contraction effect to explain the muon decay paradox, unlike the STR, which gives contradicting results if we use both it's length contraction effect and time dilation effect to explain the muon decay paradox.

4) According to our theory, the maximum relativistic mass an object can have is four times it's rest mass, while, according to the STR, there is no upper limit on the relativistic mass an object can have, since, as the speed of the object approaches the speed of light in vacuum, the relativistic mass of the object approaches infinity. This means, according to the STR, one should be able to make an electron have a mass equal to that of a muon or a proton, by accelerating it within our particle accelerators, while according to our theory, that can never happen. This is something that can be tested to determine the validity of our theory and the limitations of the STR.

5) From #4, above, we see an inherent contradiction within the STR. If we accelerate an electron within a particle accelerator, then according to the STR, there is no

reason as to why all the energy that is being put in, to accelerate the electron, simply not go into just increasing the mass of the electron according to the $E = MC^2$, of the STR, without increasing the kinetic energy of the electron. In other words, we should have an electron with relativistic mass, but a non-relativistic kinetic energy! If we take this one step further, and start with an electron at rest, relative to the accelerator, then we will have, without violating the STR, an electron with relativistic mass, i.e. $v \neq 0$, but with zero kinetic energy, i.e. v=0! This contradiction does not arise with our theory, as the kinetic energy is independent of the relativistic mass, which means, that either all the energy put in can go towards accelerating the electron or a part of the energy can go towards increasing the mass with the rest going towards the kinetic energy of the electron. Also, since, our theory says that the maximum relativistic mass of the electron can only be four times it's rest mass, if we keep putting in energy into the electron, then once the electron achieves it's maximum relativistic mass, any more energy can only go into increasing it's kinetic energy only. The fact that in our theory, the kinetic energy does not depend on the relativistic mass, means our theory does not allow the situation where, v=0 and $v \neq 0$, at the same time, unlike what we saw above with the STR.

6) A simple experiment, that can be done, to validate our theory is as follow: let us take a photon that is traveling at the speed of light in vacuum and slow it down to rest. Even though a photon has no rest or relativistic mass, it does have momentum and we can give the photon, using this fact, an infinitesimal rest mass, ϵ_p. Then, from our relativistic mass equation, we can see that when the photon, which has been traveling at the speed of light in vacuum, is gradually slowed down to zero speed, then it must give up $\frac{3^{th}}{4}$ of it's energy, in some form, perhaps as a radiation, and it's wavelength should increase four fold. This kind of experimental scenario is not possible with the STR due to the Lorentz factor being present in the equation for the relativistic mass.

7) Finally, I like to mention that I have already written a manuscript that goes into a full and complete discussion regarding all the consequences of the existence of two

kinds of S'. One of the consequences is that the Space within our universe can not only be Euclidean, but can also be both finite and unbounded.

Relativity and the Universal Constants.

I) Introduction:

The relativistic theory we will be considering here is our relativistic theory, which is based on the invariance of Newton's second law for motion and the constancy of the speed of light in vacuum (which is available in an introductory form in my published book, "New Ideas for a New Era"). In this paper, it will be shown that, if we accept relativity as a fundamental characteristic of our Universe, then, we also have to accept that not all the so-called Universal Constants can be universally constant. It will be seen, through examples, that some of the Universal Constants are truly universally constant, while others are not. The point of this paper is to introduce the fact that before we consider something as a Universal Constant, it is imperative that we first check it due to the relativistic nature of our Cosmos.

II) Examples of some of our Universal Constants and the effect of relativity on them:

Before giving the examples of some of our Universal Constants and showing whether relativity permits them to be Universal or not, I would like to set up the stage we will be using to do our experiments.

Our stage consists of two non-absolute inertial reference frames, S and S', with S' moving at a constant speed of Γ, relative to S, along the +x-axis of S, with the +x'-axis of S' parallel to the +x-axis of S. We will also place two physicists at the origins of S and S' and call them P and P'.

a) The speed of light in vacuum, C:

Let us say that P measures the speed of a light beam in vacuum and finds it to be equal to C and P' measures the speed of the same beam of light, and finds it to be equal to C'. Since, our relativistic theory is based on the assumption that the speed of light in vacuum is the same for all inertial reference frames, we can conclude that C'=C. Thus, the speed of light in vacuum is a true Universal Constant. I am aware of the fact that we had made the speed of light in vacuum to be constant for all inertial reference frames, by definition, and hence, what we have done above is simply stated our definition in a different way. Nevertheless, we will still say that the speed of light in vacuum is a true Universal Constant.

b) The universal gravitational constant, G:

Our physicist P will write the Newton's law for gravitation, for a mass M, acting on a unit mass μ, i.e. $\mu = 1$, at a distance of r, from the center of mass of M, using the spherical coordinate system, as: $F(M,\mu,r) = -G\frac{M\mu}{r^2}$. The negative sign is to show that the force, F (M, μ, r), is directed towards the center of mass of M. Our physicist P', due to the equivalence of all non-absolute inertial reference frames, must write his equation for the Newton's law for gravitation, for the same masses, M and μ as: $F'(M',\mu',r') = -G'\frac{M'\mu'}{(r')^2}$. Here G' is the gravitational constant relative to S', and the M', r' and μ' are the corresponding values for M, r and μ, relative to S', respectively. Since, our relativistic theory took the invariability of the Newton's second law for motion as being axiomatic, we have, $\frac{dP}{dt} = \frac{dP'}{dt'}$. This leads to, F' (M', μ', r')=F (M, μ, r), or $-G'\frac{M'\mu'}{(r')^2} = -G\frac{M\mu}{r^2}$, (Equation #1). Since equation #1 has to be true for all values of the variables in it, M, M', r, r', μ and μ', we have to conclude that, G'=G. Thus, we find that the gravitational constant is a true Universal Constant.

c) The Planck constant, h:

If we take a particle traveling along the +x/x'-axis at uniform speed v/v', relative to S and S' respectively, our physicists P and P' will write the De Broglie equation for this particle as $\lambda = \frac{h}{mv}$ and $\lambda' = \frac{h'}{m'v'}$, respectively. From our relativistic theory, we have, $l' = \left(1 - \frac{\Gamma}{c}\right)l$, where l' and l is the length of a rod along the +x'/x-axis, relative to S' and S, respectively. Since, wavelength is just a distance or length, similar to l and l', we can substitute λ and λ' for l and l' in the equation between l' and l. This gives us, $\lambda' = (1 - \frac{\Gamma}{c})\lambda$. From this we get, $\frac{h'}{m'v'} = (1 - \frac{\Gamma}{c})\frac{h}{mv}$ (equation #2). Applying the same argument to equation #2 as for equation #1, and bearing in mind that here Γ is not a variable similar to the other variables, but is a constant between S and S', we get $h' = \left(1 - \frac{\Gamma}{c}\right)h$. (Equation#3). Based on our relativistic theory we can also write the equation #3 as, $h' = \left(1 - \frac{v}{c}\right)h$, where v is the speed of an object relative to S. We can immediately see that for non-relativistic conditions, i.e. where, $\Gamma \ll c$ or

$v \ll c$, equation #3 becomes, h'=h. Here we have a supposedly Universal Constant that we find is not universally constant! This is a profound conclusion with immense consequences. Some of the consequences of the non-universality of the Planck constant are:

1) We cannot use the same value for h when studying non-terrestrial phenomenon. This means the value of h at the core of the Sun, where nuclear fusion is occurring, has a different value than here on Earth.

2) When studying the radiation from the accretion disc at the center of our Milky Way, due to a supposed existence of a supermassive black hole, we need to use a very different value for h, since the speed of the material, v, is very close to C.

3) From, $h' = \left(1 - \frac{\Gamma}{c}\right) h$, or $h' = \left(1 - \frac{v}{c}\right) h$, we see that for $\Gamma \simeq c$ or $v \simeq c$, $h' \simeq 0$.

4) Depending upon the way the atomic clocks are built to study the time dilation effect of Einstein's Special Theory of Relativity, it is possible that a good part of the effect noted is due to the difference between h and h' between the "stationary" and "moving" clocks, respectively.

5) For light, we have $v = c$, which gives us h'=0 and $\lambda' = \frac{0}{0}$, obtained from $\lambda' = \frac{h'}{m' v'}$, and m'=0 for a photon. This means, light can have an infinite number of wavelengths. We have, $h' = \left(1 - \frac{\Gamma}{c}\right) h$ and $\lambda' = (1 - \frac{\Gamma}{c}) \lambda$. For S' of the first kind, we have, $h' = \left(1 - \frac{v}{c}\right) h$ and $\lambda' = (1 - \frac{v}{c}) \lambda$. For light, with v=c, gives us, $h' = 0$ and $\lambda' = 0$. We get the energy, E of a photon, relative to S, $E = \frac{h' c}{\lambda'} = \frac{0}{0}$. However, if we use S' of the second kind, which was described in our relativistic theory, we have $h' = \left(1 - \frac{\Gamma}{c}\right) h$ and $\lambda' = (1 - \frac{\Gamma}{c}) \lambda$ and $\Gamma \leq c$. For a photon we have v' = c, relative to S' and v = c, relative to S. With this we get E for the photon, relative to S, to be given by, $E = \frac{h' c}{\lambda'} = \frac{(1 - \frac{\Gamma}{c}) h c}{(1 - \frac{\Gamma}{c}) \lambda} = \frac{h c}{\lambda}$. Thus, instead of $E = \frac{0}{0}$, obtained using S' of the first kind, we get a proper equation for E, relative to S, using S' of the second kind. This reinforces the importance of being explicit regarding which kind of S' one is using to describe or explain a particular phenomenon.

6) From the consequence #3 above, we should observe the wave-particle duality breaking down in our particle accelerators as $v \to c$, as this makes $h' \to 0$. This leads to $\lambda' \to 0$, since, for particles, $m' > 0$ (m' being the relativistic mass of a particle, relative to S). Unlike

Einstein's special theory of relativity, our relativistic theory makes $m_o \to 4m_o$ as $v \to c$, and not infinity.

7) It is also easy to see the breaking down of Heisenberg's uncertainty principles as $v/\Gamma \to$ c, making $h' \to 0$.

III) <u>Conclusion</u>:

As seen in the preceding paper, we find that one must be careful in taking a constant to have the same value universally. We found the gravitational constant to have the same value universally, while the value of the Planck's constant depended on the inertial reference frame. Since, relativity, is a fundamental characteristic of our Universe, it means that before we use the same value for a constant for any phenomenon within our Universe, we should make sure whether our assumption is valid or not in the first place.

<u>A short essay based on physics, on Galilean, Newtonian, Einsteinian and other kinds of Times and their consequences.</u>

I) Introduction:

Time is an abstract entity that we all feel and have tried throughout history to measure it. Due to the abstract nature of Time, and more for the sake of simplicity than due to any physical proof, we had assumed that it must be of only one kind and have invented various devices to measure it. I will not go into the various ways before Galileo that we humans have attempted to capture Time in one of our inventions. Galileo showed how to measure Time using the periodicity of a pendulum with great success, as we all know, and which we will call, Galilean Time (GT). Isaac Newton later introduced a Universal Time associated with an Absolute spatial frame of reference (ARF), which he labeled it as 't' in his equations and which we will call Newtonian Time (NT) and which is The Universal Time. Einstein showed that, if there is an ARF and a Universal Time, they cannot be determined and hence we cannot concern ourselves with something that is beyond measurable. Each non-ARF (whether inertial or non-inertial) has it's own time, which we will call, Einsteinian Time (ET). Since, there is no Absolute Space and Absolute Time, according to Einstein, the relative space, and time, relative to a non-ARF, forms one unit, called the space-time continuum. The space and time of a non-ARF are smooth or continuous as opposed to being quantized and hence the word 'continuum' attached to the unit called space-time of a non-ARF. In his Special Theory of Relativity (STR) he showed how the times of two Relative Frames of Reference (RRF) S and S' are related. In addition, in his General Theory of Relativity (GTR), which is really a theory of gravitation, he showed how the gravitational field of an object affects ET due to the curvature of the space-time continuum (STC) around the object. Thus, the ET within a gravitational field (GF) of an object can be considered another kind of time, the Gravitational Einsteinian Time (GET). Thus, so far we have seen several kinds of Times, namely, GT, NT, ET and GET. I maintain that there is no a priori reason to discard one kind of Time in favor of another. The physical proof of the existence of one kind of Time does not necessarily and sufficiently disproves the continued existence of the other kinds of Time. In fact, I maintain that we need different kinds of Time to explain different physical phenomenon within our Universe. In the next section, we will see what we can derive and learn about these Times and find out other kinds of Times. By studying the relationships between these different kinds of Time, we will learn some quite fascinating aspects of our Universe and physical phenomenon within it.

II) Since GT depends on gravity and we also have GET that is dependent on gravity, we see that on the surface of an object, that is not a Black Hole (BH) or the Event Horizon (EH) of a BH, $\Delta GT >$ ΔGET (inequality #1). As one moves away from the surface, ΔGT decreases and ΔGET increases, eventually, there will be a distance $'r_E'$, from the center of the object, which we will assume to be spherical, where $\Delta GT (r_E) = \Delta GET(r_E), i.e \ GT = GET$. For $r > r_E$ we have $\Delta GT < \Delta GET$. For $GF = 0$, $\Delta GT = \infty$ and $\Delta GET = \Delta ET = finite$ and $GET = ET$. At the EH of a BH, $\Delta GET = \infty$ and $\Delta GT = 0$. This is clearly the opposite of inequality #1. Even on the surface of a BH, a BH being a singularity actually does not have a surface, but for the sake of argument we will assume that a BH has an infinitesimally small surface), SBH, we still have $\Delta GT = 0$ and $\Delta GET = \infty$. Outside the EH, there is a distance, $r_E(EH)$, from the EH, where, $\Delta GT\{r_E(EH)\} = \Delta GET\{r_E(EH)\}, i.e \ GT = GET$. Since NT is independent of gravity, we have $NT(r) \neq 0$ for all r of a spherical object and $NT \neq \infty$ for a BH/EH either. Thus NT is always finite and same everywhere within our Universe, including at the EH and the "surface" of BH. Since, $\Delta GET = \infty$ at the EH and SBH, we can conclude that there is a $r_N(BH)$ outside the EH, where $\Delta NT = \Delta GET$ or $NT = GET$. Similarly, there is a r_G for a planet, or an object that is not a BH, where, $\Delta NT = \Delta GT$ or $NT = GT$. These r's, namely, r_E, $r_E(BH)$, $r_N(BH)$ and r_G are all real, which means, GT and NT are as real as ET and GET. Therefore, the existence and reality of ET and GET is neither necessary nor sufficient to make GT and NT unreal/imaginary and non-existent. Another result of the above discussion is that there is no one entity called, "The Time", but Time comes in multiple types/kinds/forms, which are all different from each other, except that they coincide under specific situation(s) and/or condition(s). Thus, our Universe "runs" on different kinds of Times simultaneously, which means when we need to explain a particular phenomenon, we need to use the appropriate kind of Time. So far we have found that Time comes in the forms/kinds called, NT, GT, ET, GET, we can say, unless there is a definite proof to the contrary, that there are Times that are unique to the other types of fundamental forces, besides Gravity, i.e., there is an Electrostatic Time (EST), a Magnetic Time (MT), a Weak force Time (WT), a Strong force Time (ST). Since it was shown by Maxwell that the Electrostatic force and the Magnetic force are different forms of a single Electromagnetic force (EMF), we can say that the EST and MT are also different forms of a single

Electromagnetic force Time (EMT). In addition, it has been shown that EMF and Weak force (WF) are two versions of a single force called the Electro-Weak force (EWF), we can say that EMT and WT are two versions of a single Time, the Electro-Weak Time (EWT). The hope is to unite all the forces into a single Unified force (UF), which will mean the GT, EWT and ST are different forms of a Unified Time (UT). We have seen from $NT = ET/(1 - \frac{u}{c})$ (this is from the paper on, "A relativistic theory based on invariance of Newton's second law for motion and the constancy of the speed of light in vacuum"), that at the moment of Big Bang (BB), $u = 0$, making NT=ET. In addition, at that moment there is only the UF and hence only UT. This means, at the moment of BB, NT=ET=UT and GT=GET=0, since there is no Gravitational force (GF) present yet. From NT=ET=UT, we can also conclude that, NT=UT. When the Universe evolved, ET split away from NT due to $u \neq 0$. Similarly, GT and GET also separated from NT as they are no longer the same as NT now. This means, (a) UT separated completely from NT or (b) other than ET, GT, GET, the other Times have remained as part of NT. If (a), then NT is on it's own, while the UT continued to split into the four fundamental forces Times. But, if (b), then in order to understand the three non-gravitational forces, we will need to use NT and not ET. A corollary of (b) is that, if we can determine the relationships between the different Times that are related to the four fundamental forces, then, one may truly be able to unify all of them into one Unified force/field. This should not be impossible to do, as we have shown, as an example, in our relativistic theory, the relationship between NT and ET.

Though we cannot measure NT now, we can still notice its presence/existence through the phenomenon of Quantum Entanglement. Here, even though, we have, at least, two particles that are separated in space, they are still bound by Time. The Time here is the NT. Even though, according to Einstein's Special Theory of Relativity (STR), the maximum speed with which information can possibly travel is that of the speed of light in vacuum, with respect to NT, the entire Universe, which includes us, of course, is in the present. We can get a glimpse of this from $\Delta t' = \left(\sqrt[2]{1 - \frac{v^2}{c^2}}\right)(\Delta t)$, which we all know is from STR. For light, traveling in vacuum, $v = c$, making $\Delta t' = 0$. Thus, relative to light, the entire Universe is always in the present. But, with NT, being the Absolute Time of an ARF (it has to be inertial only, as it is Absolute, i.e. $v = 0$, relative to itself and any non-ARF, S). The entire Universe, which includes not only us but also our Quantum Entangled particles, is always in the present, relative to NT. This means, the Time, t, in the wave equation describing the Quantum Entanglement of our particles should be NT and not ET. This leads to the situation where any change in one particle must cause an instantaneous change in the other particle for them to remain in the present, relative to NT. Obviously, the information here cannot be carried by light, even at its speed in vacuum, but by something else. Since, it is NT that is keeping the particles in the present at all times, it means, either, (a) NT itself is carrying the information or (2) it is a fundamental characteristic of our Universe itself due to the continued existence of NT within it even after BB. If (a), then the speed of the information has to be infinite. An immediate corollary of this is that NT, in particular, can be a carrier of information and Time, of any kind, in general, can be a carrier of information. If (b), then Quantum Entanglement proves the reality of the continued existence of NT within our Universe, even after $NT \neq ET$, i.e. when $u \neq 0$, even though we can no longer measure NT, unlike the ET, at the present stage of the evolution of our Universe. From this one cannot conclude that all Quantum Phenomenon (QP) uses NT, but other QP, for example, gravitational interaction between objects, uses other kind(s) of Time(s), including, of course, ET.

Finally, we can see that our Universe must be temporally finite and unbounded. This is due to the reality and continued existence of NT. With NT, the entire Universe is always in the present, i.e. finite, and, since, there is no upper limit to NT, it must also be unbounded, i.e. without limit and instantaneously "jumping" from one present to the next, forever. This leads to a most interesting characteristic of NT. Since, relative to NT, the entire Universe is always in the present and "jumps" from one present to the next, it means that NT is quantized and not smooth. This characteristic of NT leads to, (a) ET is quantized also, since, we have found a mathematical relationship between NT and ET and the denominator has ratio of speeds and therefore has no units to make ET non-quantized. This means ET is not smooth, which, in turn, means, the STC is quantized, at least

with regards to Time, (b) that NT is a more appropriate Time to use in studying certain QP rather than ET, such as the Quantum Entanglement. Of course, one may need to use ET, even though it is also quantized, for other QP and even non-QP. The same goes for NT also, and lastly, (c) the quantized nature of NT makes it an ideal kind of default Time to use in Quantum Physics, unless one finds that NT is not sufficient to provide a mathematically consistent explanation of a QP.

A mathematical proof of the equivalence of Dark Matter and Dark Energy using the Lambda Theorem.

(I) Introduction:

In my book, "New Ideas for a New Era", it was proposed that Dark Matter (DM) and Dark Energy (DE) are equivalent. However, a formal proof of this equivalency was not given. The entity representing DM/DE, which has a gravitational effect on matter and which pervades the entire Cosmos, was given the symbol Λ. Here, the equivalency of DM and DE will be proved using The Lambda Theorem.

(II) The Lambda Theorem:

The most general formulation of the theorem is as follows: Given an object with mass, M, volume, V, outer surface area, S, the Lambda Theorem, in spherical coordinates with the origin at the center of mass of the object, is given by: $\Lambda(R, \vartheta_i) > d^2 \Lambda_V(R, \vartheta_i) + K$, where, (a) $\Lambda(R, \vartheta_i)$ is the value of Λ on S at the point (R, ϑ_i), (b) R is the distance of (R, ϑ_i) from the origin, (c) ϑ_i are the angles made by the vector $\vec{R}$ with the three Cartesian axes, x, y and z, in that order, and whose origin coincides with that of our spherical coordinate system, (d) the values of ϑ_i are given by $0 \leq \vartheta_i \leq 2\pi$, (e) $\Lambda_V(R, \vartheta_i)$ is the value of Λ, from within the volume V, at the point (R, ϑ_i) and, lastly, (f) K is a universal constant and $K < 0$.

(III) Derivation of the Lambda Theorem:

In the interest of the reader, a simpler derivation of the Lambda Theorem will be given here. However, it should be noted for the record, that a complete, formal derivation is available in my possession.

For a simpler derivation of the Lambda Theorem, let us consider a spherical object of mass M, radius R, volume V, and outer surface area S. Let us also assume that the density of matter, ρ, of our object is constant, i.e. $\rho = constant$. Let us use a spherical coordinate system whose origin is at the center of our spherical mass. Let us consider our object in two parts. One part will consist of the outermost shell of infinitesimal thickness, while the other part will consist of the rest of the object. Let us now consider each part separately.

(a) <u>The mass without the outmost shell</u>:

Let us take an infinitesimal volume $äV$ of this mass. The mass of this volume will be given by $\delta M = \rho\,(\delta V)$. The Λ_V will act on this δM. We have assumed that the action of Λ_V is gravitational and since we are considering an infinitesimally small mass, δM, the action of Λ_V will be uniform on it. This means for every action of Λ_V on δM. There will be an equal but opposite action of Λ_V on δM, which will cancel, it's effect. This means the net action of Λ_V on δM will be zero. Extending this reasoning to the entire mass under consideration, we can conclude that the net action of Λ_V on the mass under consideration will be zero. This also means that there will be no contribution to our Lambda Theorem from the Λ_V acting on the main mass of an object.

(b) <u>The outermost shell of infinitesimal thickness</u>:

Since this is the outermost shell, it will have the outer surface S and an inner surface S_I. Also, since the thickness of the shell is infinitesimal, $S \approx S_I$.

Let us divide this shell into a collection of infinitesimally small cubes of volume δV and mass δM. Again we will have $\delta M = \rho\,(\delta V)$. The surface S will be acted upon by $\Lambda(R, \vartheta_i)$ at each point (R, ϑ_i). Therefore, the surface δS, that forms one of the surfaces of δV, will be acted upon by $\delta \Lambda(R, \vartheta_i)$. The inner surface S_I and the mass within the shell will be acted upon by an infinitesimal amount of Λ_V, i.e. by $\delta \Lambda_V$. Each of the four sides of the cube at (R, ϑ_i) that is within the shell will be acted upon by an infinitesimal amount of $\delta \Lambda_V$, i.e. by $\delta\,\{\delta \Lambda_V(R, \vartheta_i)\}$ or $\delta^2 \Lambda_V(R, \vartheta_i)$. The effect of $\delta^2 \Lambda_V(R, \vartheta_i)$ on one surface will be cancelled out by the equal and opposite effect of $\delta^2 \Lambda_V(R, \vartheta_i)$ on the opposite side. This will result in zero net effect on the part of the cube that lies within the shell. The effect on the surface of the cube formed by S_I, δS_I will be acted upon by an infinitesimal amount of $\delta^2 \Lambda_V(R, \vartheta_i)$, i.e. $\delta\,\{\delta^2 \Lambda_V(R, \vartheta_i)\}$ or $\delta^3 \Lambda_V(R, \vartheta_i)$. Even-though $S \approx S_I$, S is still slightly greater than S_I and by extension $\delta S > \delta S_I$. Therefore, the effect of $\delta \Lambda(R, \vartheta_i)$ on $äS$ will be greater than the effect of

$\delta^3 \Lambda_V(R, \vartheta_i)$ on δS_l. This can be written as, $\delta \Lambda(R, \vartheta_i) > \delta^3 \Lambda_V(R, \vartheta_i)$. Taking the limit, where $\delta \to 0$, the above inequality becomes $d\Lambda(R, \vartheta_i) > d^3 \Lambda_V(R, \vartheta_i)$. Integrating both sides of this inequality gives us $\Lambda(R, \vartheta_i) > d^2 \Lambda_V(R, \vartheta_i) + K$, where K is the integration constant. This inequality is true for all $\Lambda(R, \vartheta_i)$ and $\Lambda_V(R, \vartheta_i)$, including for $\Lambda(R, \vartheta_i) = 0$. The $\Lambda_V(R, \vartheta_i)$ is same as $\Lambda(R, \vartheta_i)$ since the subscript 'v' on $\Lambda_V(R, \vartheta_i)$ is simply to tell us that the Λ we are considering is within the volume 'V'. Therefore, when $\Lambda(R, \vartheta_i) = 0$, $d^2 \Lambda_V(R, \vartheta_i)$ must also be equal to zero. This will make our inequality 0>0+K or 0 > K. This can also be written as K < 0. With this, we have our Lambda Theorem, $\underline{\Lambda(R, \vartheta_i) > d^2 \Lambda_V(R, \vartheta_i) + K,}$ where, K < 0 and is a Universal Constant.

Re-writing our Lambda Theorem as $\{\Lambda(R, \vartheta_i) - d^2 \Lambda_V(R, \vartheta_i)\} > K$ and remembering that, (a) $\Lambda(R, \vartheta_i) \equiv \Lambda_V(R, \vartheta_i)$ and (b) $d^2 \Lambda(R, \vartheta_i)$ is an infinitesimally squared amount of $\Lambda(R, \vartheta_i)$, we can conclude that $\Lambda(R, \vartheta_i) > K$, with K < 0.

IV) Conclusions from the Lambda Theorem:

For the sake of making the arguments simple, we will use the simpler formulation of the Lambda Theorem, $\Lambda(R, \vartheta_i) > K$. The generality of the results will not be affected since we are dropping the 2nd order differential of $\Lambda(R, \vartheta_i)$ from our theorem.

(1) Let us consider the case where $\Lambda(R, \vartheta_i) < 0$ with $|\Lambda(R, \vartheta_i)| < |K|$ (inequality #1). As a reminder, we know that K < 0. To know the physical meaning of inequality #1, we need to recall that, (a) Λ has a gravitational (G) effect on matter, and (b) that, conventionally, the effect of a gravitational field on matter is given a negative sign. This is to show that the effect is a "pulling" effect. From these, we can conclude that Λ can act as a regular attracting gravitational force. When it does this, it is "pulling" the matter, upon which it is acting, towards itself.

(2) $\Lambda(R, \vartheta_i) = 0$. The physical meaning behind this equality is straightforward. It means that the net effect of Λ on an object is zero. This means, Λ can have a "repelling" or anti-gravitational (AG) effect on matter.

(3) $\Lambda(R, \vartheta_i) > 0$. In this case, we have Λ having an anti-gravitational (AG) or "repelling" effect on matter. In #2 we had already seen that this effect of Λ must

exist. This "kind" of Λ we have been calling Dark Matter (DM) and Dark Energy (DE). When this "kind" of Λ is surrounding an object, it will have an AG effect on it. This will result in the object being squeezed from all sides. This is precisely what we see with DM. If, however, this "kind" of Λ is present only on one side of an object, then its effect on the that object will be that of the object being pushed away with acceleration as per Newton's 2nd law for motion. This is, also, precisely what we see with objects that are supposedly near The Universal Boundary (TUB). This phenomenon, we have assumed, to be caused by DE. However, we see that it is the same Λ in case of both DM and DE. This proves the equivalency of DM and DE. This can also be expressed as DM≡DE.

(4) From #3, we can also arrive at the following: For objects that are near TUB and are, therefore, being pushed away, we should see some deformation of that part of the object that is facing us. This deformation is obviously due to the Λ acting as DE on the object in question.

(5) In #3 we showed that DM≡DE. Taking clue from Newton, who showed that the terrestrial and non-terrestrial gravity are the same, we can replace the terms DM and DE with the term "anti-gravity". This would mean that both DM and DE are anti-gravity.

(6) Since, we can have Λ > 0, at any time 't', the Universe at the time of the Big Bang (BB), i.e. t = 0, could not have been a singularity. The radius of the Universe at t = 0 had to be greater than zero. If we represent the size of the Universe at t = 0 by σ, then we can take this to be the size of a Quantum of Space. In the paper on the kinds of Time, it was shown that the Newtonian Time (NT) jumps from one "present" to the next. This is similar to Quantum Jumping. This means we can say that NT is quantized. We can represent this Quantum of NT as τ_N. In the paper on a relativistic theory based on Newton's 2nd law for motion, we showed the relation between the Newtonian Time and Einstein Time (ET). Therefore, if NT is quantized, ET must also be quantized. We can represent a Quantum of ET as τ_E. Therefore, not only is Time quantized, but so is Space.

(7) Firstly, if we accept that Space is quantized, as shown in #6, and the size of a Quantum of Space is σ, then, a Black Hole (BH) cannot be a singularity. The smallest

size of a BH will be equal to σ. This means, Einstein's General Theory of Relativity (GTR) cannot be complete, as it says that BH's have to be singularities. This also means, the Space-Time Continuum (STC) at the Event Horizon (EH) of a BH cannot have an infinite curvature. In fact, since neither Space nor Time is continuous, we cannot even talk about STC at the Quantum level. Secondly, one cannot have a particle whose size is smaller than σ. If we consider a particle to be a wave-packet, as per DeBroglie, then the size of the smallest wave-packet will be equal to σ.

(V) <u>Final thoughts</u>:

The Lambda Theorem does not tell us anything about the physical nature of Λ. It just tells us what kinds of gravitational effects Λ can have on matter. In section IV we have seen that using the Lambda Theorem, we can arrive at some quite profound results. The most important of these results being the equivalency of DM and DE. The author has already developed a gravitational theory based upon Quantum Interactions that shows the physical nature of Λ. This theory will be presented, G-D Willing, separately and shortly.

The Principle of Mass Equivalence.

In the Science journal, Vol. 347 no. 6226 pp. 1096-1099, there is an article by Adrian Cho, on a very sensitive experiment that is going to be done by a Stanford University group to test the "Principle of Mass Equivalence". Galileo first established this principle and later Einstein used the thought experiments of a free falling elevator and a rocket moving with a constant acceleration in empty space (i.e., where Gravitational Field =0) and concluded that for a given object, it's gravitational mass, m_G, has to be equal to it's inertial mass, m_I, i.e. $m_G = m_I$. However, using the very same thought experiments of Einstein, I found, mathematically, using logically and physically valid arguments, that $m_G = m_I + \Omega$, where, Ω is a universal constant and, $0 < \Omega \leq 1$. This means:

1) $m_G \neq m_I$, in general, but only approximately, when, $m_I \gg \Omega$.

2) Since for a photon, $(m_I)_p = 0$, we see that, $(m_G)_p = \Omega > 0$. Thus, even though a photon or light has no inertial mass it has a gravitational mass. We also see that Ω has to be extremely small. The other conclusions we can derive about photon/light from above are, (a) the bending of light due to gravity of an object is not due to curvature of the space around it, but due to the effect of gravity on the non-zero gravitational mass of a photon/light, (b) since the Ω is a universal constant, we also see that the deflection of light due to gravity has to be independent of it's frequency, (c) as the Ω has to be extremely small, since it is equal to the gravitational mass of a photon, the best way to test the equivalence principle is to use light itself. Using any other object(s), besides photon(s), might not reveal the non-equivalence of the gravitational and inertial masses, since the current instruments may not be precise enough to measure Ω. This will then lead to the false conclusion that the principle of mass equivalence is not approximately valid but exactly valid.

3) In the article, it says that the Eöt-Wash experiments have shown the equivalence principle to be accurate up to one part in 10 trillion. But, the Eöt-Wash experiments (which, the article says are considered to be the gold standard for the proof of the Principle of Mass Equivalence), and which are described briefly, did not use photons

and hence, based on #2 above, can only be said to show that the principle is approximately valid up to one part in 10 trillion.

4) If anyone is interested to see, I have the entire derivation of $m_G = m_I + \Omega$ as a hand written manuscript. Since Einstein took the principle of equivalence to be exactly correct when deriving his theory on gravitation, and we see that it is only approximately correct, and that too for $m_I \gg \Omega$, this means the GTR has to be, at most, an approximately correct theory. This may be the reason that the GTR gives rise to singularities and cannot be extended to t=0, when the Big Bang occurred. Using the correct expression between gravitational mass and the inertial mass, I have also been able to derive an equation relating time and the radius or size of the Universe. This equation allows t=0 and says that the radius or size of the Universe at that moment was not zero, i.e. the Universe was not a singularity or an infinitely dense point of matter/energy.

Using similar and valid logical, physical and mathematical arguments, it can be shown that the general relationship between the gravitational mass and inertial mass is given by, $m_G = m_I \pm \Omega$. From this equation, it can be seen that:

(1) Gravity can be either attracting or repelling. This can be more easily seen with how a photon would behave in a gravitational field. Since, for a photon, we have $(m_I)_p = 0$, this will make $(m_G)_p = \pm\Omega$. As we have seen above, $+\Omega$ results in deflection of a photon towards an object, in whose Gravitational Field the photon happens to travel through. From this, we can conclude that, $-\Omega$ results in the deflection of a photon away from an object in whose Gravitational Field it happens to pass through. This means the Gravitational Field of an object can be either attractive or repelling or attractive up to a certain distance from the object and repelling after that.

(2) The above can also be expressed as: The GF can be < 0 (attracting, which is the kind we are all familiar with) or > 0 (repelling or anti-gravity, whose existence we are beginning to be aware of through Dark Matter and Dark Energy).

(3) Using the above #2 conclusion, it is easy to see the possibility for an object of mass M to have a GF that is both attracting and repelling. In my published book,

"New Ideas for a New Era", I have shown how this is possible, without violating logic, physics or mathematics.

(4) The conclusions of #1, #2, and #3 are both profound and immense. I have worked out many of these conclusions, which are available as a hand written manuscript. A few of those I would like to present here for the interested reader: (a) It is possible to make any object invisible to the radar without using special shapes or radar absorbing paints. (b) It is possible to build vehicles that can use anti-gravity for motion. The speeds and the directions of the movements of such vehicles will be nothing less than miraculous. (c) The existence of anti-gravitational fields solves the problem of Dark Matter and Dark Energy and also shows that they have to be equivalent, given our most current knowledge of their physical effects on visible matter, (d) A vehicle surrounded by an Anti-GF is immune to any attack on it of any kind whatsoever, (e) There are many more theoretical and practical consequences, which I have already worked out but cannot go into here without making this short paper into a comprehensive research manuscript. Nevertheless, one can easily see how useful a vehicle that uses anti-gravity will be for commercial and non-commercial purposes.

A theorem on the nature of Tachyon and its consequences.

I) Introduction:

Gerald Feinberg coined the term "Tachyon" in 1967 to describe particles traveling at speeds greater than the speed of light in vacuum, C. Here, I present a theorem on two of the characteristics of a Tachyon and the consequences thereof. The theorem will be for a Tachyon symbolized by ℵ, but it should be clear that the theorem is fully general and applicable to any number of Tachyons.

II) Presentation of the theorem:

If an entity ℵ is a Tachyon, then, (1) its rest mass must be zero, i.e. $(m_0)_ℵ = 0$ and (2) the speed of ℵ in vacuum, $C_ℵ$, is the same in all inertial reference frames, i.e. $C'_ℵ = C_ℵ$.

III) Proof of the theorem:

Besides the above possibility, which is our theorem, we can have three other possibilities. These are as follows: (a) $(m_0)_ℵ \neq 0$ and $C'_ℵ = C_ℵ$, (b) $(m_0)_ℵ = 0$ and $C'_ℵ \neq C_ℵ$ and (c) $(m_0)_ℵ \neq 0$ and $C'_ℵ \neq C_ℵ$. Let us now consider each of these possibilities individually. If (a), then we can simply put $C_ℵ$ for C in our relativistic theory that is based on the invariance of Newton's 2nd law for motion. Then, from our equations $l' = \left(1 - \frac{\Gamma}{C_ℵ}\right) l$ and $m(v) = m_0 \left(1 + \frac{\Gamma}{C_ℵ}\right)^2$, we get for $v, \Gamma = C_ℵ, l' = 0$, which means the volume of ℵ is equal to zero, and $m(C_ℵ) = 4 (m_0)_ℵ$. Since $(m_0)_ℵ \neq 0$, we get the density, ñ, of ℵ equal to infinity, which is physically absurd. This eliminates the possibility (a). If (b), then we can use our relativistic theory with C' = C. In this case we get l', with $\Gamma = C_ℵ$, from our equation $l' = \left(1 - \frac{\Gamma}{C}\right) l, l' < 0$. This is also a physical absurdity. The third possibility, (c), where $(m_0)_ℵ \neq 0$ and $C'_ℵ \neq C_ℵ$, can be eliminated using the same argument as in (b) where we had C' = C. This completes the proof of our theorem as we have eliminated all the possibilities except one, which is our theorem. Hence, the conditions expressed in our theorem regarding $\aleph$ have to be true.

III) <u>Consequences from the theorem</u>:

(1) Since $(m_0)_\aleph = 0$, $\aleph$ cannot be matter.

(2) Since $C_\aleph > C$, $\aleph$ cannot be pure energy.

(3) Since $\aleph$ is neither matter nor pure energy, it cannot be directly detected by our instruments, which are based on detection of matter and/or energy only.

(4) A corollary of #3 is that none of our senses can detect $\aleph$ either.

(5) If neither our senses nor our instruments can detect $\aleph$, it is a transcendental entity. This does not mean that it is cannot be as real as matter and energy.

(6) From $C'_\aleph = C_\aleph$, we can conclude, using light as our analogy, that the speed of $\aleph$ in vacuum is always constant.

IV) <u>Final thoughts</u>:

Here we will speculate on the other characteristics of $\aleph$.

We have found that $\aleph$ is neither matter nor pure energy, but a third kind of entity that can exist in our Cosmos. If this is the case, then our law of conservation of matter/energy will need to be re-defined. The law will need to be changed to the law of conservation of matter, energy and $\aleph$. We do not know if $\aleph$ can interact with matter or pure energy. However, if $\aleph$ does exist, then we can substitute $C_\aleph$ for C, the speed of light in vacuum, in our relativistic theory that is based on the invariance of Newton's 2nd law for motion, by making $v_0 = v = C_\aleph$ when deriving ë. In this way, all our relativistic equations will have $C_\aleph$ instead of the C. This will mean that it should be possible for matter to attain supra-luminal speeds without violating any laws or resulting in physical absurdities, such as, l' < 0 or $\rho = \infty$. This is irrespective of whether or not $\aleph$ does in fact can interact with matter and/or pure energy. If $\aleph$ cannot interact with matter/pure energy, then it can have only one speed, i.e. $C_\aleph$, the speed in vacuum, in all mediums made up of matter and/or pure energy. Even though $\aleph$ may not interact with matter/energy, the fact that it has a speed, $C_\aleph$ means that it has kinetic energy (E_K). If the $C_\aleph$ is not due to the kinetic energy we are familiar with, then, it will have to be due to kinetic energy we do not yet know. This will mean the kinetic energy can come in two forms. One form is the one that propels matter, while the other form propels $\aleph$. We can represent these kinetic energies as $(E_K)_m$ and $(E_K)_\aleph$. The existence of $(E_K)_\aleph$ will again force us to change our law of conservation of matter/energy/$\aleph$ to $m_T + E_T + \aleph_T + \{(E_K)_\aleph\}_T = Constant$.

If we combine the possibility of matter able to travel at supra-luminal speeds, as shown above, with the idea, from the paper on "Mass Equivalence", that anti-gravity must exist, then it is a matter of engineering to construct a vehicle using which humans can travel not only at supra-luminal speeds but, in ways that defy our current imagination.

There are at least two ways we can indirectly notice the existence of $\aleph$. The first is by calculating $C_\aleph$. Since, in our relativistic theory we can substitute $C_\aleph$ for C, our equation, $E_m(v) = m(v)C^2$, becomes $E_m(v) = m(v)C_\aleph^2$. For v = 0, we get the equations, $E_m(0) = m(0)C^2$ and $E_m(0) = m(0)C_\aleph^2$. The former gives the expected amount of energy from m (0), while the latter gives the actual amount of energy obtained from the same m (0). Comparing the actual amount of energy from m (0), $\{E_m(0)\}_{actual}$, to the expected amount of energy from the same m (0), $\{E_m(0)\}_{expected}$, we can calculate $C_\aleph$ from the equation:

$$C_\aleph = \sqrt[2]{\frac{\{E_m(0)\}_{actual}-\{E_m(0)\}_{expected}}{m(0)}} + C^2.$$

The second way is by observing matter/particles traveling at supra-luminal speeds. This is allowed by our relativistic theory that is based on the Newton's 2nd law for motion.

Finally, the fact that we can simply substitute $C_\aleph$ for C in our relativistic theory, which is based on Newton's second law for motion, shows its generality, malleability and ultimately its superiority over Einstein's Special theory of relativity.

On a mathematical theory of the Nervous System and Consciousness.

I) Introduction:

In this paper, we will introduce the mathematical laws of the Nervous System (NS) and Consciousness (C). We will use the experimentally proven concept of Quantum Entanglement (QE) from Physics, and our Sohraab-Hyder (SH) Set Theory from my book, "New Ideas for a New Era". Here we will introduce the three laws that govern the NS and from them derive consequences, including Consciousness (C) and Self-Consciousness (SC). We will also derive other conclusions, which, I am sure, the reader will find both novel and fascinating. One of these conclusions is that the demise of the physical body, necessarily, does not lead to the demise of the Consciousness or the Self-Consciousness. This is true even if the physical body disintegrates into atoms and is spread over the entire Cosmos!

II) The Three Laws of the Nervous System (NS):

Before I state the three laws of the NS, I like to define and clarify the terms we will be using in those laws.

(a) When we use the term "set", we will always mean a Sohraab-Hyder (SH) set.

(b) For the time, t, we will mean the Newtonian Time (NT) as was discussed in my paper, "A short essay on the Galilean, Newtonian, Einsteinian and other kinds of Time".

(c) For a Quantum Mechanically Entangled Process (QMEP), i, we will use the symbol, Q_i.

(d) To show the dependency of Q_i on time, t, we will use the symbol, $Q_i(t)$.

(e) For the $Q_i(t)$ that is part of Consciousness (C) and Self-Consciousness (SC), we will use the symbols, $Q^C_i(t)$ and $Q^{SC}_i(t)$, respectively.

(f) For the set of $Q_i(t)$, we will use the symbol, $\{Q_i(t)\}$. It is assumed that the set $\{Q_i(t)\}$ is always finite.

(g) We will use the other symbols of the set theory with their usual meanings, such as, $\subset$ for subset and so forth.

(1) <u>The 1st law of the NS</u>:

The 1st law of the NS states that $NS = \{Q_i\}$.

(2) <u>The 2nd law of the NS</u>:

The 2nd law of NS states that, $Q_i = Q_i(t)$.

(3) <u>The 3rd law of the NS</u>:

For a Nervous System Function (NSF), i, $(NSF)_i$, we always have, $(NSF)_i \subseteq NS$ and $(NSF)_i = \sum_1^n a_j\,(t)Q_j(t)$. Here, (a) $(NSF)_i = \{Q_j(t)\}$, (b) $a_j(t)$ is the fractional contribution of $Q_j(t)$ to $(NSF)_i$ at time 't'.

III) <u>Some conclusions that can be obtained from considering the above three laws of the NS</u>:

(1) From the 3rd law, we can see that for any $(NSF)_i$, $\sum_1^n a_j\,(t) = 1$.

(2) The $(NSF)_i$ applies to all of the NS functions, which means, it applies to (a) objective, (b) subjective and (c) both objective and subjective. An example of (a) is a hand movement, of (b) is an emotion/feeling, including the feeling of being Conscious and Self-Conscious, and an example of (c) is a behavior with an emotion behind it, such as speech.

(3) A corollary of #2, based on the 3rd law is that all the types of NSF are governed by $\{Q_i(t)\}$.

(4) If we put $(NSF)_i = NS$ in the 3rd law, we can conclude that $NS\,(t) = \sum_1^n a_j\,(t)Q_j(t)$.

(5) Since, $(NSF)_i \subseteq NS \Rightarrow NS = \{(NSF)_i\}$.

(6) If $NS\,(t) = \sum_1^n a_j\,(t)Q_j(t)$ and $NS\,(t) = \sum_1^n b_j\,(t)Q_j(t)$ for any 't', then since, $NS\,(t) = NS\,(t)$, for all 't', $\Rightarrow \sum_1^n a_j\,(t)Q_j(t) = \sum_1^n b_j\,(t)Q_j(t)$ (equation #1). From this equation #1, we get, $\sum_1^n (a_j - b_j)\,(t)Q_j(t) = 0$ (equation #2). Since, $Q_j(t) \neq Q_i(t)$ in equation #2, we get, $(a_j - b_j)(t) = 0$, or $a_j(t) = b_j(t)$. This means, at any time 't', the equation, $NS\,(t) = \sum_1^n a_j\,(t)Q_j(t)$ is unique.

(7) From the 2nd law we can conclude that since 't' is quantized and is always in the "present", this means NS (t), at all times 't' is a unitary system.

(8) From #2, we can say that Consciousness is a type of (NSF)$_i$ and according to our 1st law must be a part of NS. We will give the (NSF)$_i$ that represents Consciousness the symbol C (t) and define it by the set $\{Q^C{}_i(t)\}$. From the 3rd law we can also say that C (t) $\subseteq NS$ (t) and C $(t) = \sum_1^n a_j$ $(t)Q_j(t)$.

(9) Again, from #8 above and the 3rd and 1st laws, we also will have Self-Consciousness, which we will give the symbol SC (t). Hence, SC $(t) = \{Q^{SC}{}_i(t)\}$, SC (t) $\subseteq NS$ (t) and SC $(t) = \sum_1^n a_j$ $(t)Q^{SC}{}_j(t)$.

(10) As both C (t) and SC (t) are two types of (NSF)$_i(t)$ and one cannot talk about SC (t) without the necessary existence of C (t), we must conclude that SC (t) $\subseteq$ C (t). This also means SC is a special kind of C, just as C is a special kind of (NSF)$_i$.

(11) From C $(t) = \{Q^C{}_i(t)\}$, we get the set { } or a null set. This null set obviously has to be the set representing Unconsciousness or UC (t). Therefore, UC (t) = { } and UC (t) = 0.

(12) A corollary of #11 is that since there can be only a single null set, any NS (t), at any time 't', must have an UC state, (b) UC must be unique, i.e. there cannot be more than one type of UC state.

(13) From the 1st law we have $NS \equiv Q_i(t)$, which means NS can be, (a) localized, (b) non-localized or (c) partially localized and partially non-localized. These possibilities are due to the nature of $Q_i(t)$. This means, if at $t = t_A$ the NS (t_A) is localized, then, it is not physically impossible for the NS at $t = t_D$ to become non-localized, due to some physical process on the localized NS, and still remain as a unitary system.

(14) A corollary of #13 is that C (t) can be localized at t_A and non-localized at t_D. The same goes for SC (t). This is the physical and mathematical basis for the possibility of the continuation of Consciousness and Self-Consciousness even after the demise of the physical body. Religions have called this possibility as the continuation of the Consciousness after death or as "life after death".

IV) Final thoughts:

In section III, we derived some results using our three laws for the Nervous System. However, it is clear, that what we had derived is just the tip of an iceberg. There are many more results obtainable from our three laws, as well as from the combinations of the results already obtained. From $Q^C_i(t)$ and $Q^{SC}_i(t)$, it can be seen that both Consciousness and Self-Consciousness can be of different kinds and since they are also functions of time, they can change with time. For example, at certain time(s), we can have C (t)=SC (t), while at other time(s) C (t) ≠ SC (t). A similar situation can arise with Unconsciousness, where UC (t) = C (t) at certain time(s), while at other time(s) UC (t) ≠ C (t). The neurological condition called "vegetative state" can be expressed in our mathematical language as, {SC (t)=UC (t)} ≠C (t). This expression is possible because both SC (t) and UC (t) are ⊆ C (t).

From our result #14, we can see the possibility of a non-localized SC (t) becoming partially or wholly part of a localized SC (t). This coincidence can be sporadic or continuous for a certain length of time. This forms the basis of the strange phenomenon where some people seem to be in contact with a person who no longer physically exists as a localized physical being.

From the equality NS (t) = SC (t), we can conclude that at certain time(s) a person can feel being Self-Conscious throughout his/her entire being. These time(s) are when one feels "alive throughout the body".

Finally, for the question whether there is any experimental proof for $NS = \{Q_i(t)\}$, I have the following: (a) From our 1st and 2nd laws of the NS, the equality is valid. (b) We know that the NSF's are physical processes and hence come under the purview of Quantum Mechanics. (c) $Q_i(t)$ are, therefore, Quantum Mechanical Processes. (d) $Q_i(t)$ is an experimentally proven fact.

<u>A novel method to evaluate large amounts of data on chronic neurological diseases and its consequences.</u>

ABSTRACT:
         The purpose of this article is to view chronic neurological diseases using a powerful and novel mathematical method. Relapsing-remitting multiple sclerosis was used as an example to show the novelty and the power of this mathematical method/tool. This method reduces all the available information about a chronic neurological disease to a mathematical function that fully describes the current knowledge for that disease. It is then shown how this function can be used to predict the course of the disease in a given patient and how it can be used to stratify the treatments for that disease. Finally, it is shown how the mathematical function can be used to compare different treatments without the need for doing long and expensive head to head clinical trials.

I) <u>Introduction</u>: Thanks to the painstaking work by large number of people, there is now available an enormous amount of information on various chronic neurological diseases. This information is usually presented as a 2-dimensional Cartesian coordinate-time mapping consisting of a variable 'x' against the time 't'. Another way is to compare one variable ' x_1 ' against another variable 'x_2 '. One of the important limitations of these ways of looking at a disease is that, from the limited information obtainable from 2-dimensional mapping contradictory results about the entire disease itself can be obtained. It is not uncommon, to see conclusions, which are supposed to be applicable to the entire disease, reached from, x v/s t, contradict conclusions reached from, x_1 v/s x_2. A method to look at the entire data, with as many variables as possible, at once, would be much better and less likely to give results that are likely to be unrealistic or inapplicable to the entire disease.

II) The method I propose is to take as many variables 'x' about any disease 'D' and map them on a (n+1) dimensional space with time as one of the variables. Time is a very important variable to include, since one is dealing with chronic conditions. Hence, for any patient 'i' one can map all the ${}_D^n x_i$ variables known against 't'. Using this information, a function f_i^D (${}_D^n x_i$, t) can be obtained for each patient 'i'. From the collection of these f_i^D (${}_D^n x_i$, t), a "generic" function F_D (${}_D^n x$, t), can be derived, that is applicable to the entire disease 'D'. This function will be the most accurate and realistic description of the disease 'D' given the variable ${}_D^n x$. With time, as more information is found, i.e. more ${}_D^n x$ are found, the function F_D can be further fine-tuned. This function F_D, I like to call the Khan (or K) function for the disease 'D'.

III) For the purposes of illustrating the power and utility of this method, I like to apply it to one of my favorite chronic neurological disease, the relapsing-remitting multiple sclerosis (RRMS). It will be immediately clear that this application to one disease is not unique, but is fully general and applicable to any chronic disease. In the case of RRMS, the ${}_R^n x$ will refer to the EDSS, MSFS, number of T_2 lesions supra-tentorially, number of T_2 lesions infra-tentorially, number of T_2 lesions within the spinal cord and so forth. The greater the number of variables considered at one time, the better, which is the opposite of the other methods, such as the 2-dimensional method. Then for each patient 'i' one obtains f_i^R (${}_R^n x_i$, t) and from these we get the K function for RRMS, F_R (${}_R^n x$, t). This then gives the full description of RRMS as it is currently known. From this, one can derive certain very useful consequences:

a) The first one is that, one may find that the F_R (${}_R^n x$, t) is not unique. It may be that the spread of f_i^R (${}_R^n x_i$, t) is such that one has to have several F_R (${}_R^n x$, t), suggesting several sub-types of RRMS (i.e. RRMS1, RRMS2, RRMS3...). This would not have been possible to find without the use of this methodology. Finding the sub-types of RRMS, in turn, has great consequences for the research and treatment of this disease.

b) The second consequence is that when a patient 'i' is first given the diagnosis of RRMS (i.e. at time t_0), one can predict what the ${}_R^n x_i$ for this patient will be at any other time t_1. For this prediction, we will need a "closeness" (or "distance") quantity

C_t which defines how "close" this patient is to F_R ($_R^n x$, t). For the present, I would

recommend to define $C_t = \sqrt[2]{\Sigma_t(_R^n x - _R^n x_i)^2}$ at time 't'. Knowing C_{t_0}, one should be

able to extrapolate or obtain the f_i^R ($_R^n x_i$, t) for this patient under the assumption

that the f_i^R ($_R^n x_i$, t) remains parallel to F_R ($_R^n x$, t) and hence obtain $_R^n x_i$ for any future

time 't'. The assumption of f_i^R ($_R^n x_i$, t) being parallel to F_R ($_R^n x$, t) is not unreasonable

given that F_R ($_R^n x$, t) defines RRMS. This information should help a lot in the ensuing

discussion with the patient, such as, what course the disease, in time, will take in

that particular patient.

c) The third conclusion is, that for any treatment 'T' for RRMS, one should be able to

obtain a K function F_R^T from the longitudinal data of a clinical treatment trial. By

comparing the C_t between the F_R and F_R^T we can immediately stratify the efficacy of

the various treatments. An immediate corollary of this stratification is that by

comparing the C_t between f_i^R ($_R^n x_i$, t) and F_R^T we can decide at t_0, which treatment

will be most appropriate for the patient 'i'. Another corollary of the stratification of

the efficacy of the various treatments is the non-necessity of doing head-to-head

clinical trials of the various treatments, which that tend to be both long and quite

expensive.

IV) <u>Final thoughts</u>: Considering the above discussion, on using the proposed

method for RRMS, it should be quite clear that this method is not specific for RRMS,

but can be applied to any chronic neurological disease and even to chronic non-

neurological diseases. In addition, this methodology would have been nearly

impossible to implement before we had the modern computers and the computer

graphics. However, with the current computer technology, this methodology is quite

easy to implement, and places in our hands a powerful mathematical tool which

should provide, through visualization, an understanding of any chronic disease in a

unique and powerful way that had never been considered before.

About the Author

I am a neurologist by profession. However, ever since I was fourteen years old, I have been passionate about mathematics and physics. I won the annual prize for mathematics when I graduated from high school and my nickname in high school was "Newton", given to me by my fellow students. I did my bachelor in science in physics, with highest distinction and was awarded the annual Professor Abel's Memorial Award for physics. After obtaining my degree in physics, I went into medicine and became a neurologist. I am a member of the National Physics Honor Society. I am also a member of the American Mensa Society. During my neurological training and as a practicing neurologist, physics remained as a hobby with me. I tried to keep up with the latest news in physics through the television programs and the popular science magazines, to the extent that was possible, given my extremely busy professional and family life. I always had many ideas in my mind regarding various branches of physics, especially cosmology, but I did not have the time to put them on paper and develop them into papers worthy of publication. By the Will of G-D, I was able to find some time now to develop some of those ideas and publish them as a book for the wider audience. I continue to practice as a neurologist and I live with my wife, Nadia, and two sons, Sohraab and Hyder in the State of New York.

Printed in the United States.
D/E technicians

Printed in the United States
By Bookmasters